MW00437847

Water Power for the Farm and Country Home

WATER POWER

FOR THE

FARM AND COUNTRY HOM

By DAVID R. COOPER

STATE OF NEW YORK
STATE WATER SUPPLY COMMISSION
APRIL, 1911

STATE OF NEW YORK

STATE WATER SUPPLY COMMISSION

HENRY H. PERSONS, President

MILO M. ACKER,

CHARLES DAVIS,

JOHN A. SLEICHER,

ROBERT H. FULLER,

COMMISSIONERS

DAVID R. COOPER,

Engineer-Secretary

WALTER McCULLOH,

Consulting Engineer

LYON BLOCK, ALBANY, N. Y.

Water Power

FOR THE

Farm and Country Home

ι Compliments of

CONSERVATION COMMISSION

State of New York

[Into which the former State Water Supply Commission has been merged]

Second Edition

PRINTED FOR THE STATE WATER SUPPLY COMMISSION
BY J B LYON COMPANY, STATE PRINTERS
ALBANY

ROBERT H. FULLER,
COMMISSIONERS

DAVID R. COOPER,
Engineer-Secretary.

WALTER MCCULLOH,
Consulting Engineer.

LYON BLOCK, ALBANY, N. Y

Water Power

FOR THE

Farm and Country Home

BY DAVID R. COOPER

Engineer-Secretary

New York State Water Supply Commission

Second Edition

PRINTED FOR THE STATE WATER SUPPLY COMMISSION

BY J B LYON COMPANY, STATE PRINTERS

ALBANY

WATER POWER FOR THE FARM AND COUNTRY HOME

BY DAVID R COOPER

In the course of its general investigations of the water powers of the State, the Water Supply Commission has heretofore confined its attention to the possibilities for large developments, and the regulation of the flow of rivers and large creeks No previous or general investigation of small creeks and brooks and their power possibilities has been made, not because they were considered unimportant, but because the Commission believes that if the State decides to take an active part in the regulation of the flow of streams and the development and conservation of water powers, it should confine its first activities to the larger units, leaving the smaller opportunities for later examination and for private and individual development However, no comprehensive system of conservation can meet with universal favor unless it contemplates the prevention of waste, great or small, and wherever found.

Accordingly, the Commission desires to call attention to the valuable power which is now running to waste in thousands of small creeks and brooks in all sections of the State. Many of these minor streams present possibilities for small individual developments of power sufficient to supply all the requirements of the owner at a comparatively small cost Numerous farms in the State have on them brooks or creeks capable of supplying power sufficient to furnish electric light for all the buildings. Others would also furnish power enough to drive a feed grinder, a churn or cream separator, or to run a wood saw, sewing machine or other machines and implements requiring a small amount of power for their operation In short, there are numerous small streams now tumbling over ledges in barnyards or pastures whose wasted energy might readily be transformed and applied to useful work by the installation of small and inexpensive water-power plants. If the power of more of these were developed and substituted for manual labor, a great saving of time and energy would be accomplished, and financial profit would result.

After the initial expense of installing the plant is paid, the cost of a small water power is inconsiderable, the plant requiring little personal attention and small expense for supplies and repairs. However, while the power of some streams may be developed at an amazingly small cost, in other instances the

Modern Application of Hydro-electric Power Vacuum Milking Machines

cost may be prohibitive. In this connection, one fact that is perhaps not fully appreciated is that the power of a waterfall is comparatively permanent, only its rate of availability being limited. While the stream may shrink in the dry summer and fall, it is quite certain to swell again in the spring and to continue the process, year after year, as the source of supply is continually renewed. But the power which might have been, but was not developed in the year 1910, cannot be reclaimed in 1911 or ever after. Much of the power that is wasted by inequality of the flow of the stream may be saved by conservation through water storage; but this sometimes involves a large outlay and therefore, generally speaking, the fullest use of the power of a small stream can best be obtained by using the stream as it runs, or at best after temporary storage behind inexpensive dams.

The Water Supply Commission believes that the possibilities for small water powers should be pointed out to the people of the State in order that there may be a better realization of the usefulness and value of this remarkable natural resource and that the farmers and residents of rural districts may take advantage of the opportunities to conserve and utilize them. It is believed that some facts relating to the utility of power in general and small water powers in particular, together with descriptions of some typical small water-power developments that are now in actual operation, and brief notes as to how such a power may be developed and applied,will suffice to bring the subject forcibly to the attention of those most interested, and furnish at least a beginning for observations in this comparatively new field, and stimulate a tendency to a more general utilization of this source of power, and a consequent saving of much energy now secured from coal, wood and other exhaustible producers of power. Accordingly, the following discussion of the many and varied uses for power on the modern farm, together with descriptions of developments now in use, and notes on developing a small water power, are submitted in the hope that they may be of interest and service to those who have chosen farming for their livelihood or pleasure, especially by assisting them in the consideration as to whether or not it may be worth while to develop the power of any particular stream. These discussions and descriptions are not intended to suffice as a practical handbook for laying out a power plant, but merely to point the way to an intelligent consideration of the possibilities, by showing what others have done and laying down a few fundamental principles, which should properly be taken into consideration in determining upon the development of a small water power.

"Luminous" Electric Radiator

USES FOR POWER ON THE FARM

The impossibility of securing a sufficient number of capable and satisfactory farm hands in these days, when the majority of young men are turning to the populous centers for their

Motor Lifting a Ton of Hay, Hydro-electric Power

livelihood, is perhaps the most compelling reason why machines which can be substituted for manual labor are a decided advantage to the up-to-date farmer. Their adoption as a part of the permanent equipment for the farm should render their owner comparatively independent of some of the problems of supply and demand for farm labor, the solution of which problems is an important factor in determining the success or failure of the farmer who disposes of his produce in open market. This condition is supplemented by a commendable tendency for farmers to live better, to place the home life of the farm on a higher plane, and to make farming a means of pleasurable livelihood rather than the mere eking out of a bare subsistence from the products of the soil. These conditions, together with the greatly improved quality of illumination and convenience which electricity affords, are creating a growing demand for a reliable and reasonably economical source of energy with which to supply both light and power on the larger estates and farms.

Electric Toaster

That electric light is much cleaner and more convenient than kerosene lamps must, of course, be admitted by all. It must also be admitted that a kerosene lamp of any considerable illuminating power has also certain heating propensities which render it an unpleasant companion on a warm summer evening. However, when it comes to a consideration of mere dollars and cents, there seems to be a widespread belief that kerosene as a source of illumination is cheaper than electricity. Statements to this effect are too often allowed to go uncontradicted, and too many people accept this view without taking the trouble to investigate.

It is a comparatively simple matter to compare the cost of the two kinds of light, knowing as we do exactly how much current an electric lamp of a certain filament and candlepower will consume. Such a comparison will frequently result in a choice of electricity as the cheaper light. In many cases the selection of electricity to supplant kerosene lamps would result in no considerable saving of money, but would do away with considerable inconvenience and furnish much better illumination. If cost is the controlling consideration, the comparison cannot always be so much in favor of electricity. An important consideration, often overlooked, is that with electric lights the interiors of living rooms do not require such frequent repapering or refinishing as they would require with kerosene illumination.

However, the convenience and cleanliness of electricity are fairly well known and appreciated, but the means by which electric currents may be generated economically, and by which this form of energy may be applied to bring about sufficient returns, financial and otherwise, to warrant the installation of an isolated plant for a farm or country home, are not so generally understood.

Motor-driven Sewing Machine

Electric current may be generated by means of a dynamo, or generator, with any kind of a power-producing plant. All that the dynamo requires to enable it to produce electric current is power of some kind that may be applied in such manner and quantity as will cause the armature, or "interior core," of the machine to rotate at a sufficiently high and uniform rate of speed. There are various kinds of power generators which will perform this work satisfactorily for isolated plants. Within the last few years the small internal combustion engine, supplemented by the electric storage battery, for stationary service, have been so much improved and simplified as to cause them to compare very favorably with the better-known types of power-producing apparatus in first cost and in reliability of operation. The extreme simplicity of both this type of engine and of the storage battery, together with the great economy in fuel consumption of these engines, the low price of fuel, and the efficiency of the battery as a device for storing the energy and delivering it in the form of electric current when needed and in the quantity required, result in a low operating cost. The advent of tax-free alcohol into the field of available fuels for use in internal combustion engines, and the growing demand for this class of fuel, indicate that it will become, in time, a strong competitor of kerosene and gasolene At present, gasolene is the fuel most generally used for engines of this type and small-size gas engines are now manufactured by many firms

Steam power is probably the best understood of all classes of power. In many cases, especially where the fuel is very cheap, this is the best power for a farmer to have Steam-power plants, as well as gasolene, kerosene and alcohol plants, all require personal attendance during operation and necessitate more or less frequent applications of fuel Wind power is also a source of energy which may well be considered by the farmer who needs a small amount of power

Perhaps the most promising source of power for farmers in New York State is the power that may be developed from falling water This kind of a power plant requires comparatively little personal attention while in operation, and needs no replenishing of fuel except such as Nature herself provides in the flowing brook Not only are there many of these powers that are undeveloped as yet, but there are many others which have been developed at some previous time and have recently been allowed to fall into disuse for various reasons Many old sawmills were abandoned when the surrounding hills were all lumbered off A small investment would enable many such old powers to be revived and applied to some useful purpose Such a water-power plant could frequently be made to serve the owner or a group of users of electric current at very small first cost for each individual, and at an operating cost which would be inconsiderable

It should be borne in mind, however, that much depends on the choice of the best power for any particular purpose, and a careful consideration of what is needed, and the conditions under which the power must be supplied, is essential to insure satisfaction with a power plant In any particular instance a manufacturer of small waterwheels will cheerfully submit an estimate for a water-power plant, while the makers of steam and gasolene engines will quite as readily furnish any information to be based on data furnished by the intending purchaser.

The extent of the applications of power to practical purposes on the farm is very broad. While perhaps electric lighting is the use most frequently thought of, it is, however, in the application of electric current or power to the operation of labor-saving devices that the greatest gain is to be derived on the large farm or country place. Feed grinders, root cutters, fodder cutters, fanning mills, grindstones, circular saws, corn shellers, drill presses, ensilage cutters and elevators, horse clippers, milking machines, grain separators, threshing machines, cream separators, churns, vacuum cleaners, ice cream freezers, dough mixers, feed mixers, chicken hatchers, and numerous other machines and implements operated by power, are obtainable in these days of labor-saving devices. The amount of power required to operate many of these is small. The presence of a plant of sufficient capacity to operate one or two particular machines often makes it possible to use the power for many of the other purposes. The amount of work that a small power will do may be judged from the following brief statements of what is actually being done:

Motor-driven Ice Cream Freezer

Six horsepower will drive a grain separator and thresh 2500 bushels of oats in ten hours.

Three horsepower furnishes all power needed to make 6000 pounds of milk into cheese in one day.

Six horsepower will run a feed mill grinding twenty bushels of corn an hour.

Motor-driven Cream Separator
Note small size of motor

Five horsepower grinds twenty-five to forty bushels of feed, or ten to twelve bushels of ear corn, an hour.

Seven horsepower drives an eighteen-inch separator, burr mill and corn and cob crusher and corn sheller, grinding from twelve to fifteen bushels of feed an hour, and five to eight bushels of good, fine meal.

Six horsepower runs a heavy apple grater, grinding and pressing 200 to 250 bushels of apples an hour.

Five horsepower will drive a thirty-inch circular saw, sawing from fifty to seventy-five cords of stovewood from hard oak in ten hours.

Six horsepower saws all the wood four men can pile in cords.

Twelve horsepower will drive a fifty-inch circular saw, sawing 4000 feet of oak, or 5000 feet of poplar, in a day.

Electric Ironing

Ten horsepower will run a sixteen-inch ensilage cutter and blower, and elevate the ensilage into a silo thirty feet high at the rate of seven tons per hour.

One horsepower will pump water from a well of ordinary depth in sufficient quantity to supply an ordinary farmhouse and all the buildings with water for all the ordinary uses.

In determining the size of power plant required in any particular instance the use requiring the largest amount of power must be considered. It follows that there will then be plenty of power for the smaller requirements. In considering a water power it should also be borne in mind that the full theoretical amount of a water power can never be realized, a certain portion being taken up in friction in the waterwheel and in losses in the electric generator, transmission lines, motors, etc. The question as to how much may be made available will be discussed hereinafter.

Following are descriptions of some typical water-power developments in use in this State at the present time.

FARM WATER-POWER DEVELOPMENT IN ONEIDA COUNTY

On the outskirts of the village of Oriskany Falls, in Oneida county, N. Y., is a farm of about 100 acres, belonging to Mr. E. Burdette Miner. This community was at one time one of the principal hop-raising districts of the State. Mr. Miner has been engaged in raising hops for fifty years, and raised 10,000 pounds of hops on seven acres the past season. In recent years he has divided his attention between mixed farming and dairying, keeping from twenty to twenty-five cows.

Electric Hot Plate

Before the installation of his water power, not the least of the irksome tasks about the farmhouse was the daily filling and cleaning of kerosene lamps and lanterns; and the wood was sawed, and the cream separator and churn in the dairy room were operated, by hand. Five sons contributed in no small measure to the prompt disposal of the daily tasks. But the boys went forth into the world and acquired lines of activity and interest of their own. Only the oldest son remained to live on the farm. Another son studied electrical engineering, a third chose mechanical pursuits, a fourth became a civil engineer, and a fifth took up commercial work.

After coming in touch with the outer world and the great modern achievements of science and invention, especially of a mechanical or engineering character, the boys quite naturally set their wits to work to devise some way in which the daily labors of those at home might be made less burdensome.

Electric Coffee Percolator

Through the farm flows Oriskany creek, which ripples over its gravelly bed in a channel from twenty to thirty feet wide. The boys said to their father, "Why not harness the creek and make it do some of the work?" There was no precipitous fall of the creek on the farm, but the boys proposed to concentrate at least a portion of the fall by constructing a dam. This they intended to do primarily for the purpose of developing enough power to light the homestead and farm buildings with electricity and to saw the wood and do away with some of the other tiresome farm tasks.

The elder Miner was not enthusiastic at first, but was finally persuaded by the boys, who made surveys and plans for a water-power development, and in October, 1905, with the assistance of three of his boys and two day laborers, Mr. Miner began the construction of a dam across the creek. This was to be no ordinary structure. The creek, while peaceful enough

Dam of E. B. Miner, Oriskany Falls, N. Y.
Main dam at left; flood spillway at right

at most times, had a habit, well known to Mr. Miner, of bursting its bounds every spring and rushing through the farm in a torrent. So the dam was built in such a way that, while it would raise the water to a certain height during periods of ordinary flow, it would not cause the floods to rise perceptibly higher than before the dam was built. Accordingly, it was designed so that a part of it could be lowered at flood times to allow free passage for the swollen stream.

The bed of the stream at the site selected for the dam is composed of solidly packed gravel. It was not considered advisable to lay timbers on such a foundation, so a ditch about two feet deep and one and one-half feet wide was dug across the creek bed and filled with concrete, to which a heavy timber was securely bolted, to form the upstream sill for the superstructure. The downstream side was supported on a sill of heavy timber whose ends were embedded in the concrete walls, or abutments, at either end of the dam and whose middle portion was supported by posts, spaced six feet apart, which in turn rested on large blocks of concrete placed in the bed of the creek. This downstream sill was about two and one-half

Farm Power House on Oriskany Creek
Dam in left background; tail-race in right foreground

feet higher than the upstream sill. A horizontal floor of double plank extending twelve feet downstream from the upstream sill and supported by the concrete foundations under the downstream sill formed an apron for the water to fall on. This prevents back-washing under the dam. A double layer of heavy plank was then fastened on the two sills, forming a sloping face on the water side of the dam. On the upper edge of this plank-facing, at the crest of the dam, are placed flashboards, one foot high and extending the full length of the dam, thirty-six feet, but divided into six sections, each six feet long. Each of these sections is hinged by the lower edge to the crest of the dam, while the upper edge is held from tipping over by chains fastened to cast-iron lugs located about halfway down the planking The chain is held in these lugs by pins which are connected by rod and chain to a capstan, or spindle, located at one end of the dam, and are so arranged that by turning the spindle the pins will be drawn successively, thereby letting the flashboards down one at a time The idea of this arrangement is that, when a flood is rising, the capstan may be turned with a heavy lever crank, winding up the chain and pulling down the flashboards one at a time, to give more space for the flood to pass through so as to prevent the water upstream from the dam from rising too high This plan has prevented the washing away of Mr Miner's power house on several occasions

The sloping face of the dam receives the direct pressure of the water and transfers it to the sills, which in turn transfer it to the concrete foundation The reason for sloping the upstream face of the dam is that the pressure of water is always normal, or perpendicular, to the surface against which it presses, therefore, if the face of the dam is sloping, the pressure is downward, rather than outward, as would be the case with a vertical face This results in greater stability for the dam, due to the lessened tendency to tip over With a dam of this type the higher the water rises against or over it, the more nearly vertical is the line of pressure, and the dam is held tightly down on its foundation instead of tending to tip over It follows that the flatter the face of the dam the more stable it will be Mr Miner's dam raised the water about four feet

But in spite of his provision for floods, Mr Miner did not want to be under the necessity of letting down his dam for every freshet, so he provided an additional permanent spillway This is a simple concrete barrier, or wall, which flanks one end of the dam. In plan it was built at an angle with the dam proper, and extends downstream along the side of the natural bank It was built with its crest a few inches higher than the main dam, so that during periods of ordinary flow the surplus water all passes over the main dam, but as soon as the creek rises a few inches over the main dam, water begins to flow over this extra spillway, which, being about forty feet long, will discharge a considerable volume although the water flowing over it is only a few inches in depth.

This spillway is strengthened on the downstream end by a concrete abutment, which consists of a simple heavy block of concrete extending above the top of the spillway A similar abutment flanks the upstream end and also constitutes an abutment for one end of the main dam The other end of the main dam is set against the opposite bank of the creek and is protected from washing and is strengthened by a similar concrete abutment

It was considered desirable to place the little power house away from the main channel of the stream, so an earth embankment was built, extending from the downstream end of the flood spillway, a distance of about sixty feet This embankment, or dyke, is curved in such manner as to divert the water behind it across a low place to a safe distance from the main channel. Some excavating had to be done behind this embankment in order to secure a channel of sufficient depth to prevent the water from freezing to the bottom and to provide a smooth channel of approach to the power house This diversion of the water to one side from the main channel prevents the accumulation of debris and silt, which is a hindrance to the proper operation of a waterwheel. The pool thus formed is called a "forebay" and is very quiet water. The velocity of the water flowing through it is so slight that it will not carry much debris

At the downstream end of the forebay the diverting embankment approaches a steep bank At this point Mr Miner built a small power house Under the power house is the wheel-box, which consists of a box-like compartment having one side open to the forebay This opening is covered with a coarse screen to prevent leaves or other debris from entering the wheel, but the water flows through it readily. In the wheel-box a water-wheel, of the type known as a turbine, was placed This revolves on a vertical shaft, or axle, which is guided by bearings in a metal case surrounding the wheel and resting on the bottom of the box-like compartment. The wheel-case is open at the bottom to allow the free escape of the water after it has passed through the wheel. The construction of the turbines is such that the pressure of the water on the curved vanes causes the wheel to revolve, just as the pressure of wind causes a windmill to revolve. The water must have a free escape from the opening in the bottom of the wheel-case and wheel-pit and to provide for this a channel, called a "tail-race," was excavated to carry the water back to the creek. Natural conditions were favorable here and a tail-race joining the main channel about 100 feet below the power house was constructed with little difficulty. At the point where the tail-race joins the creek the elevation is two feet lower than the power house, so that there is little tendency for water to back up from the creek into the tail-race. There is a certain amount of back-water during freshets but the increased height of the water in the forebay at such times partially offsets it.

The vertical shaft of the turbine extends up through and about two feet above the floor near one end of the power house,

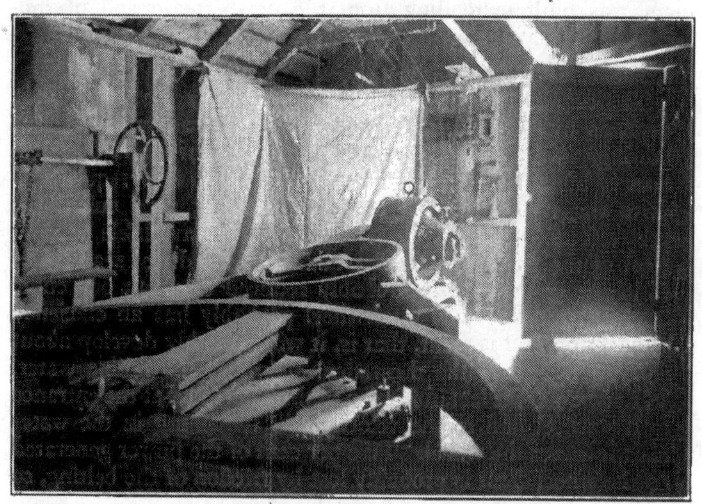

Interior of E. B. Miner's Power House

where it is supported on ball-bearings which enable it to be revolved with very little friction.

At the other end of the power house, which is twelve feet by sixteen feet in plan and seven feet high to the eaves, was placed an electric generator, or dynamo, rated at 12½ kilowatts, which is equivalent to about 17 horsepower. This machine is intended to operate at about 1100 revolutions per minute. The waterwheel, under the pressure of about six feet, would not revolve at such a high rate of speed. It was, therefore, impracticable to connect the generator shaft directly to the waterwheel shaft and it became necessary to magnify the revolutions by connecting the two shafts by belt, using different-sized pulleys.

A large wooden pulley, seventy-six inches in diameter, was keyed on the end of the waterwheel shaft A much smaller pulley, about eight inches in diameter, was placed on the driving shaft of the generator A leather belt connects the two, and since the wheel shaft is vertical and the generator shaft is horizontal, it is necessary to pass the belt over an intermediate pulley, or "idler" This idler is set with its axis at an angle with both the horizontal and vertical, so that the transition of the belt from the horizontal to vertical is made gradually Since the driving pulley on the generator shaft is so much smaller than the pulley on the wheel shaft, there are about nine revolutions of the generator shaft for every revolution of the wheel shaft

The amount of power which this equipment will generate depends to a considerable extent upon the amount of water flowing Oriskany creek at this point has a tributary drainage area of about fourteen square miles, and the flow required to drive the turbine to full capacity is about 2900 cubic feet per minute. This volume is probably available during most of the year, but is not available in the driest seasons, at which times the flow is probably reduced to about 600 cubic feet per minute The waterwheel probably has an efficiency of about eighty per cent, that is, it will probably develop about eighty per cent of the theoretical energy of the falling water The remainder is lost in friction in the wheel-box at the entrance to the wheel and in the velocity still remaining in the water after it leaves the wheel Five per cent of the power generated on the wheel shaft is probably lost by friction of the belting, so that, at rated load, about seventy-six per cent of the theoretical power of the water is probably delivered to the shaft of the generator

Mr Miner realized that there would be times when he would not require all or any of the power which would be produced. At the same time the pond formed by the dam was not large enough to store any considerable amount of water, and he had all the power he would require at any one time, so it was not considered necessary to provide storage batteries to store the electricity On the other hand he did not wish to be compelled to turn the water on and off at frequent intervals, as would be necessary unless some auxiliary regulating apparatus were provided Therefore, it was decided to provide for the plant to run continuously and to devise some means to consume the electric current when not in use. A series of resistance coils were mounted on a frame in the power house, and connected with the generator When the demand for electric current is less than the capacity of the generator, a small electric device automatically throws one or more of these coils into the circuit, and the surplus current is converted into heat by the resistance of the coils By means of this arrangement it was planned to run the plant continuously, so that whenever electric current was wanted it could be had simply by turning a switch at the house or barns

The power plant, including the dam and all the features thus far described, was completed and in operation before Christmas of the year in which the construction was begun

We have thus far seen how Mr. Miner developed his water power and transformed it into electricity It remains to see how he gets it to his house and farm buildings, and how he uses it after he gets it there

The power house is situated about 1700 feet from the house, where the electric current was most wanted This necessitated the construction of a transmission line For this purpose a double line of bare aluminum wire was stretched on a row of poles about twenty feet high and about one hundred feet apart The poles are provided with ordinary crossarms at the top on which are mounted the insulators carrying the wires. As the transmission line leaves the power house it crosses a highway and runs in a perfectly straight line to the house. Over the highway insulated wires were used as a safety precaution, but bare aluminum wire was used for the remainder because it was cheaper

The buildings are all in a cluster and a branch from the transmission line runs into each one where the current is used All the wires which are inside of any of the buildings, or are close to the woodwork, are covered with insulation, and, where concealed, are further protected by being placed in twisted metal tubes

The first actual use of this hydro-electric power was for lighting. The house was illuminated with electric lights, as were also the barn and other buildings, there being ultimately about seventy 16-candle-power lamps in use Even the pig sty has its electric light, and there is no more groping in the dark anywhere about the Miner farm buildings

But there was more power in the creek than was necessary to run the electric lights. A circular saw was brought into use, belted to a motor, and the supply of firewood was cut in a fraction of the time previously required. The same motor is used to drive a lathe and a drill in a machine shop which the Miner boys built and equipped. This motor is belted to a countershaft from which additional machine tools can be driven. One of the Miner boys has developed this machine shop as a combined means of pleasure and profit. In addition to a considerable amount of experimental machine work, he does all the farm repairs and a considerable amount of machine work for neighboring knitting mills, as well as general and automobile repair work, all of which has been made possible by the harnessing of the creek.

Lathe in E. B. Miner's Machine Shop

Another motor, two horsepower, driven by the electric current, is belted to a vacuum pump, which is connected with a one-inch pipe running to the house and the barn. In the house there are two taps, one on each floor, to which the hose of a vacuum cleaner may be attached, and Oriskany creek does the rest; the floors are cleaned in the most modern, sanitary and thorough manner. In the barn the pipe from the vacuum pump runs above the cow stanchions with a tap at alternate stanchions. The tubes of the milking machines are attached and the creek milks twenty or twenty-five cows twice each day.

In the dairy room is a one-half-horsepower motor, which may be belted to the cream separator or churn, and on the hot summer days it is frequently belted to the ice cream freezer. An ingenious float device in the separator turns off the power when the cream is all separated from the milk and trips a can of clear water into the heavy, revolving bowl of the separator, which still retains enough momentum to rinse itself thoroughly before coming to rest.

Drill in E. B. Miner's Machine Shop
Note the electric motor in background belted to countershaft near the ceiling

In a similar manner other applications of the power have followed from time to time, and one at a time most of the hand cranks on the Miner farm have been relegated to the scrap heap; even the grindstone is operated by a long, narrow belt running from the little motor in the dairy out through the door to an adjoining compartment.

In the Miner residence are five electrical heaters, which Mr. Miner states will raise the temperature to 75 degrees when it is zero outside. Since these heaters were installed there has not been much use for the wood saw. There are also in the house some electric fans which stir up a breeze on the hot days. An electric ventilator fan in the attic insures good ventilation at all times. In the kitchen the Miners cook for a family of from five to ten with an electric range, and iron with an electric iron attached by a cord to an ordinary electric lamp socket. A smaller motor operates the egg beater and cream whipper; another small motor drives the sewing machine.

E. B. Miner's Dairy Room

Vacuum milking machines in background; also small motor which drives the cream separator and churn in the foreground

The little motor in the dairy room also drives a single-acting plunger pump, which forces water up to a galvanized iron tank in the attic of the house, whence water is piped and furnished by gravity to the bathroom and kitchen. An electric heater in the kitchen heats the water for the bath and kitchen.

Other miscellaneous uses are made of the never-failing power of the creek, such as filling the silo, and the power plant requires practically no attention. Self-oiling devices on the waterwheel and generator, and the use of the resistance coils to consume the superfluous electricity, obviate the necessity for attention, except to fill the oil cups every few weeks. Practically no trouble has been experienced in the operation, the only interruption so far being due to the formation of anchor ice in the forebay, which causes a little trouble on extremely cold days. The water-wheel is run continuously, night and day, summer and winter, and electric light or current is always available at the touch of a button or by throwing a switch.

Electric Cooking Outfit, E. B. Miner's Home

As to the cost of his plant Mr. Miner would give no figures. His motto seems to be, "Not how cheap, but how good," and he states that it would require several times the cost to induce him to give up his water-power plant. Engineers estimate the cost of reproducing his plant, including the dam, power house, waterwheel, generator and transmission line, at about $1800.

SUMMER HOME POWER PLANT, NORTHWEST BAY, LAKE GEORGE

Among the attractive summer homes on the shores of Lake George is that of Mr. Stephen Loines of Brooklyn, located at the upper end of Northwest bay, about four miles above Bolton Landing. On his property there was a small lake known as Wing pond, having an area of about seven acres and situated at an elevation of about 180 feet above Lake George. The outlet was a small brook, which runs through Mr. Loines' property and flows into Northwest bay.

In the summer of 1902, Mr. Loines built a dam across the outlet of Wing pond, raising its surface about two feet. He ran a galvanized iron pipe line from the dam, down the side of the hill and along the brook. It was four inches in diameter for a short distance, then reduced to three inches and finally to two inches, and was about 1200 feet long in all, with a fall of about 110 feet. A twenty-four-inch waterwheel of the impulse type was installed in a small power house to which the

Dam at Outlet of Wing Pond

pipe line was run. The waterwheel developed about three horsepower and was belted to an electric generator.

The power was found to be insufficient to supply Mr. Loines' needs at that time. He desired to burn thirty-five 16-candle-power carbon filament lamps and to charge a 40-cell battery for an electric launch.

Accordingly, in the fall of 1908, Mr. Loines raised his dam two feet higher and installed a six-inch spiral riveted steel pipe line, running from the dam down a gulley on the surface of the ground, for about 1600 feet, to a point a short distance from the place where the creek flows into Lake George. At this point he built a small power house and installed a twenty-four-inch waterwheel of the impulse type. This wheel operates under a head of 165 feet and is directly connected by a shaft to a six and one-half kilowatt generator, which operates at 500 revolutions per minute. This generator supplies a 60-cell house battery (45 lamps), an 84-cell battery for a 35-foot cabin launch, a 48-cell battery for a 20-foot open launch and a 40-cell battery for an electric roadster, all of which are in pretty continuous use from about the first of June to the first of November of each year.

As this new development superseded the older one and proved entirely adequate for the needs of Mr. Loines' country place, the old development was made over so that it could be

Power Transmission Line, Northwest Bay, Lake George

utilized for sawing firewood to supply the superintendent's cottage and the other buildings during the winter. A countershaft was erected on the wall of the old power house, which is a building 7 feet by 10 feet in plan and about 8 feet high. This countershaft has three counterpulleys, by means of which the speed of the waterwheel may be doubled or trebled. For the purpose of sawing firewood a leather belt is placed on one of the pulleys of the countershaft and run through a small aperture in the side of the power house to the driving pulley of a circular saw, which stands on a small porch at one end of the power-house building.

Mr. Loines' superintendent stated that by operating the saw continuously for eight hours it would be possible to saw twelve cords of wood, which he estimated to be sufficient to supply his cottage, and such other of the buildings as need wood, for the entire winter. This illustrates very aptly the large amount of work that a small power is capable of doing in a short time.

In addition to lighting his house and buildings by means of the power developed at his new power house, Mr. Loines also has a rather unusual application of power on his summer place. He is an enthusiastic student of astronomy and has built a small but elaborately equipped observatory on the hillside above the cottage. The observatory is so constructed that the roof can be removed entirely from the building to a

Stephen Loines' Power House, Northwest Bay, Lake George
At left, 4-in. water pipe; at right, transmission line connection

support at the back of the observatory. The roof is mounted on wheels and Mr. Loines uses his electric power to do the work of moving the roof when he wishes to make astronomical observations with his telescope. This is accomplished by means of a small 1½-horsepower motor which operates at 1275 revolutions per minute and is connected by belt to a countershaft, which in turn is connected by a worm gear and a chain drive to the carriage on which the roof is supported. In this manner the roof may be moved the required distance in two or three minutes by simply throwing the switch which is inside the observatory building.

Mr Loines' new power house is a stone masonry building, the masonry being uncoursed rubble, constructed in a very artistic and attractive manner. The building is 9½ feet by 15½ feet in plan and is about 9 feet high to the eaves. It has a concrete foundation and the floor is of first-class concrete A concrete foundation, about 3 feet by 5 feet, provides a permanent support for the water motor and the generator This foundation projects 6 inches above the level of the concrete floor On one end of the foundation stands the waterwheel, there being an opening about 8 inches by 18 inches through the concrete base under the water motor to carry off the water after it has passed through the wheel The supply pipe for the waterwheel enters the side of the building on a level about one foot above the floor. Just inside, the pipe reduces to a diameter of about 2½ inches and is fitted with a gate valve by means of which the water may be turned on or off The nozzle of the waterwheel is also equipped with an adjusting device by means of which the size of the jet issuing from the nozzle may be varied in order to secure various speeds or the maximum efficiency of the waterwheel The setting required to give the desired speed is determined by experiment by the operator

FARM POWER DEVELOPMENT IN SCHOHARIE COUNTY

At the entrance to the driveway approach to the farmhouse of Jared Van Wagenen, Jr , at Lawyersville, Schoharie county, N Y., stand two large, stone gateway posts On the capstone of one of these posts is engraved, "Agriculture the Oldest Occupation," and on the other, "Agriculture the Greatest Science." In keeping with the latter sentiment, Mr. Van Wagenen has conducted his agricultural operations in such a manner that he is looked upon as one of the most scientific and progressive agriculturists in the State. He takes an active interest in such affairs as farmers' institutes and is considered an authority on the science of agriculture His farm and buildings are equipped with the most modern conveniences and labor-saving devices.

There is a small stream which runs through the farm and flows into the Cobleskill This stream is so small that one may easily step across it in the summer-time. About half a mile from the farmhouse is an old mill dam which forms a pond with an area of more than an acre. The dam was built long ago when small sawmills dotted that section of the State The timber having been practically all cut off, this mill, along with hundreds of others, was long since abandoned. Mr Van Wagenen conceived the idea of harnessing its wasting energy and making it do some of his farm work for him The story of how he accomplished this is best given in his own words, as follows

"About eight years ago I began to figure on how to get this power to the house where it could do a little work My first thought was to carry it there by belt cables, but figures proved that the friction would eat up the five horsepower available Electric power, easily transmitted with little loss. was the only solution. I talked with many who understood electricity and its engineering features and most of them laughed at the idea of such a small installation Had I wanted to construct a million-dollar plant there would have been whole libraries of advice, but a small plant to run entirely alone and be controlled by a seven-hundred-foot wire was evidently a novelty After a good deal of studying and feeling my way the plans were made and the work begun

"The stream being so small, the most rigid economy of water had to be observed, so I installed a nine-inch upright turbine in an upright wooden case, building the case myself, where it would get the most benefit of the fifteen-foot head This turbine, furnishing about five horsepower, I belted to a three-kilowatt, or four-horsepower, one hundred and twenty-five volt direct current generator, which would easily take care of seventy-five metal filament incandescent lamps I next installed a waterwheel governor to insure a steady flow of electricity. It took about seventy-four hundred feet of weatherproof copper wire, strung on wooden poles, which were cut on the farm, to carry the electricity to my home and the farm buildings and to the house of a neighbor As it is more than half a mile from the house to the plant it is out of the question to go there every night and morning to stop and start the machinery. Of course it is possible to let this plant run night and day during the wet season, but in dry times it is best to save the water when the power is not needed. A neighbor living about seven hundred feet from the power station kindly starts and stops the machinery with a wire stationed at his bedroom window This wire controls a valve and counterweight At five o'clock in the morning he pulls the wire and the lights come on and at a certain hour of the night he releases the wire and they go out In payment for this service I light his house and barns free of charge.

"Our maintenance charges are very small, almost negligible I think our waterwheel behaves better every year Carbon brushes for the generator last a long while and oil is a very small item Each year I am improving the plant, and very soon I expect to install a motor-driven washing machine and wringer to prepare the clothes for the electric iron and to put a vacuum cleaning outfit in the house.

"Although I consider the cost of our plant about $500, it was installed under the most rigid economy in every respect and mainly by my own hands The dam was already built and needed only some trifling repairs The gate control is my own get-up, and, while the cost is trifling, it took considerable study to get it to work right I did most of the house wiring, using concealed knob and tube for the living-rooms of the house, moulding and open wiring for the other rooms and for the barns This material cost me about $40 Of course, I do not in any instance figure in my own labor, as the work was all done at odd times"

This small power development, using the dam already built, cost Mr Van Wagenen about $500 as follows

Dynamo, 3 k w (second-hand)	$50
Waterwheel, 4 h p (naked wheel)	55
Governor (new)	75
Wire (7400 feet)	210
Labor (installing waterwheel)	40
Fixtures (lamps and the like)	38
One small motor, 2 h p. (new)	50
Total	$518

The plant furnishes power sufficient to light the farmhouse and all of the buildings with electricity, as well as those of the neighbor who turns the water on and off In the dairy a small electric motor of about 3 horsepower, actuated by the electric current, drives the cream separator and also furnishes power for running the grindstone, feed cutters, hay fork and fanning mill, in addition to which the power is also used to milk the cows and cut the ensilage and to do numerous other bits of work about the place. Mr. Van Wagenen states that his water power does work equivalent to that of a hired man the year round and does away with numerous chores and laborious duties about the place.

The arrangement which Mr. Van Wagenen devised to turn on the water at his plant and to shut it off again is unique and interesting. It consists of a triangular frame lever about two feet wide and seven feet high, hinged at one of the bottom corners. The other bottom corner is connected to a sliding gate which fits over the feed pipe for the water-wheel. At the top are fastened two wires, one of which runs to the house of Mr. Van Wagenen's accommodating neighbor, and the other runs over a pulley and has a counterweight attached to it. When the water is to be turned on, the neighbor pulls the wire and the gate is raised by the leverage of the frame; when the water is to be shut off, he releases the wire and the counterweight pulls the lever back, allowing the gate to fall in place again.

Washing Machine, Driven by Electric Motor

OTHER SMALL POWER DEVELOPMENTS

Mr. John T. McDonald, who has a farm about five miles from Delhi, Delaware county, N. Y., some ten years ago began making good use of a power development from a small stream on his farm. He lights his house and buildings, runs saws, grinders and various machines in a little shop on rainy days and in the winter His dam was made from stone and earth from the nearby fields and cost very little It forms a pond, covering, when full, about four and one-half acres of land The pond is well stocked with trout and other fish, and each winter Mr. McDonald cuts about 500 tons of ice from it Mr. McDonald turns on the water at his dam by means of an electric switch at the house and regulates the voltage also in a similar manner From the pond the water is led through a hydraulic race, or canal, about 900 feet long, to one of the farm buildings where the waterwheels are installed The head, or fall, at this point is about 15 feet and there are three waterwheels of the turbine type. one that develops 25 horsepower, another that develops 6 horsepower and a third that develops about 3 horsepower The large wheel is used to run a sawmill and feed mill The 6-horsepower wheel drives an electric generator, or dynamo, which furnishes the electric lights, and also electricity for driving the small motors about the place The 3-horsepower wheel runs the small saws, machine tools, etc , in Mr. McDonald's shop.

A few miles east from Mr. Van Wagenen's farm in Schoharie county is another small power development owned by Mr. Frank Caspar He has installed two waterwheels on a small creek and uses the power from them to drive the machinery in a table and furniture factory He has another small waterwheel of the turbine type driving a little dynamo which generates electricity for electric light Mr Caspar lights his factory buildings, his home, a neighboring church and the main street in the village with electricity from this little dynamo An ingenious device of his own invention makes it possible to start and stop the power from the house by simply pulling a wire which operates a valve in a small water pipe, from which water under pressure is let into a hydraulic cylinder This causes the piston of the cylinder to rise, and the piston being directly connected to a gate in the water-pipe inlet, allows the water to flow into the waterwheel. When it is desired to stop the plant, a pull on the companion wire causes the reverse operation to take place and the power is shut off.

Near the village of Berlin, in eastern Rensselaer county, N Y, there is a small power development owned by Mr. Arthur Cowee His source of power is a small trout brook which flows through the farm Mr Cowee is a producer of

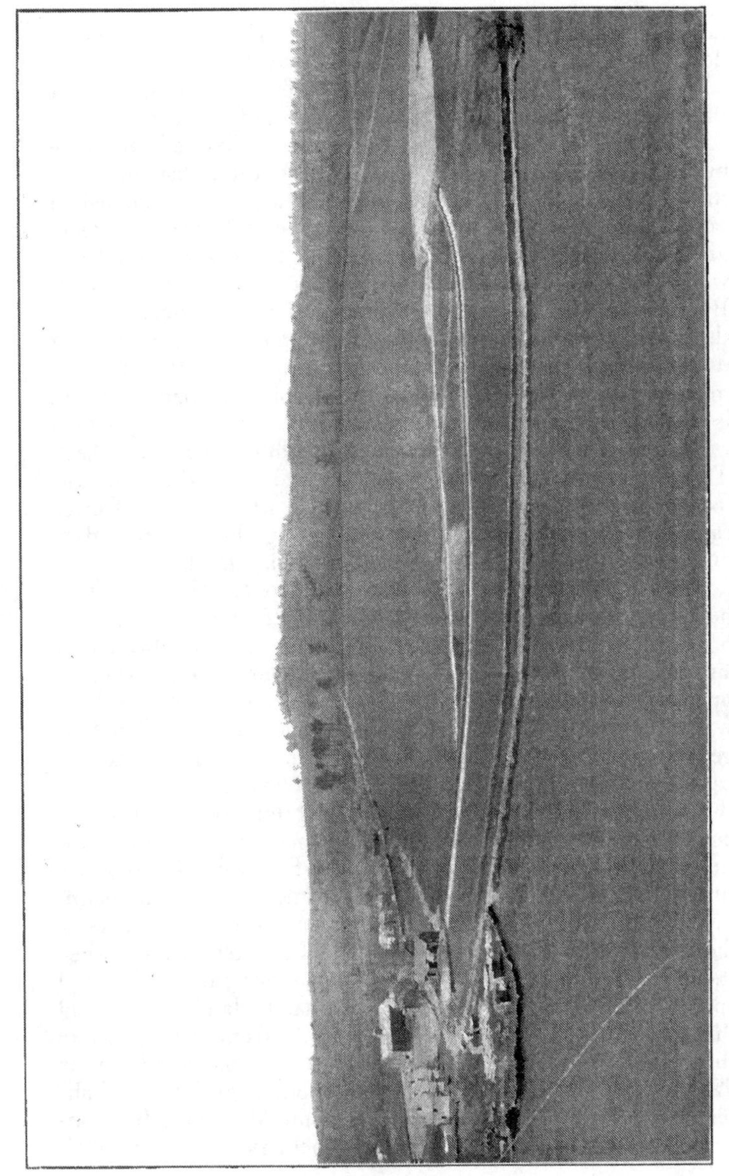

Farm Power Development of John T. McDonald, Delaware Co., N. Y.

fancy gladiolus bulbs, on a large scale His principal power development, consisting of a 36-inch impulse waterwheel, under a pressure due to a fall of about 210 feet, is used mostly for the purpose of operating a circular saw and other machinery connected with a sawmill The water is diverted from the natural channel of the brook at a considerable distance from the place where the waterwheel is installed and is carried in an artificial channel, about four feet wide and three feet deep, around the side of the hill, where it runs into a shallow basin which has been excavated by Mr Cowee at a suitable location By means of this basin, or artificial pond, practically all of the flow of the brook may be stored during the night and used to operate the waterwheel during the day In this manner the full power value of the brook is realized. There is a ten-inch, cast-iron pipe line, about 1680 feet long, which runs from the pond down the side of the hill to the waterwheel This pipe line was placed under ground from three to four feet in order to avoid freezing in the winter Mr Cowee estimates that the development, including the diverting dam and canal, pond, pipe line, waterwheel, circular saw and accessories, cost him a total of about $7000. He states that he can saw about 4000 feet of lumber in a day with this power

In addition to this development, Mr Cowee also has a small impulse waterwheel in his bulb house This wheel is operated by water furnished from the system of the local water company It is directly connected to a small electric generator which furnishes electricity sufficient for 157 sixteen-candle-power carbon-filament lamps which are installed throughout the bulb house. The generator does not produce enough electric current to run all of these lights at the same time, but it will operate as many as forty-five or fifty lights at one time, which is all that is necessary to meet the requirements

Mr. D F. Paine of Wadhams, Essex county, N Y, has a dam at the outlet of Lincoln pond. The water surface, when the pond is full, is about twelve feet above the normal and spreads over an extensive tract of low, marshy land The pond thus formed is about three miles long and from one-quarter to three-quarters of a mile wide The water is conducted from the dam to the penstock, a distance of about a mile and a half, securing a fall of 320 feet At this point Mr. Paine has constructed a power house, where he generates electricity which he transmits to Mineville for use in the mines. This power is transmitted a distance of about eight miles.

At Chazy, N. Y., near the western shore of Lake Champlain and at a point about fifteen miles north of the city of Plattsburg, there is located a modern stock and dairy farm which, in its operation, exemplifies the manifold advantages to be derived from the use of hydro-electric power for electric lighting and for the various power requirements of the farm. This farm, which is owned by Mr. W. H. Miner and is called "Heart's Delight," covers an area of 5160 acres. About 1200 acres are cultivated, 1200 acres are in pasture and the remainder in woodland. The output consists of live stock

Power House, "Heart's Delight" Farm

and dairy products, all crops grown on the farm being fed to the stock and only finished products being shipped out. The live stock includes registered Percheron and Belgian horses, pure-bred, short-horn Durham and Guernsey cattle, Dorset sheep and high-grade hogs for the production of sausage, hams and bacon. There are also poultry and squabs, and a fish hatchery for the propagation of trout. The entire output goes directly to high-grade hotels in New York, Washington and Chicago.

Two streams pass through the southern portion of the farm, the smaller one being known as Tracy brook and the larger one as Chazy river. It was decided to provide the farm with electricity for light and power. Enough water power was found in these streams to furnish a cheap and reliable source of energy. Accordingly, a hydro-electric plant was installed several years ago and has given such satisfaction that the equipment has been increased from time to time, and some novel applications have resulted. Three small concrete dams were built across Tracy brook to form storage reservoirs. A concrete penstock, or pipe, 44 inches in diameter and 670 feet long, carries the water from the downstream

Alternating Current Transmission Line, "Heart's Delight" Farm

reservoir to a concrete power house, where a fall of 19 feet is secured.

The power-house equipment consists of two water turbines automatically governed and directly connected respectively to one 30-kilowatt and one 12½-kilowatt, 220-volt, direct-current generators. The current is transmitted over a pole line, a mile and a quarter long, to a central station in the main group of farm buildings.

Another dam was built across the Chazy river. This is of concrete, and, after passing through screens at the intake gate house, built into the dam, the water flows through a concrete penstock, 48 inches wide by 60 inches high and 630 feet long, to the power house where a fall of 30 feet is obtained. There are two turbines here, belt connected to alternating-current generators, and the current is transmitted over a pole line, nearly three miles long, to the central station.

Electric Cooking Outfit

An auxiliary to the water-power development consists of two hydraulic rams, pumping water from one of the Tracy brook reservoirs to a 60,000-gallon tank, 100 feet above the ground, for fire protection for the buildings.

There are in all about twenty-five motors installed in the various buildings. The electric current actuates these motors, which are used to drive or operate numerous machines and labor-saving devices.

An entire load of hay is lifted from the wagon and stored in the mow by a ten-horsepower motor. A root-cutting machine is operated by a two-horsepower motor mounted on the ceiling. A one and one-half horsepower motor drives a vacuum pump, which operates the milking machines; five machines are used, each of which will milk two cows simultaneously. A one and one-half horsepower motor runs the cream separator, and a three-horsepower motor drives the big churn; and motors are used for driving the water pumps, as well as the brine-circulating

Motor-driven Vacuum Pump
For milking machines and vacuum cleaners

pumps in the ice-making plant A grist mill, driven by electric motor, is part of the farm equipment, and the sausage-chopping and mixing machines are driven by a four-horsepower motor Roots for the sheep are cut by a machine driven by motors of one and one-half and two horsepower, and food for the fish is prepared by a grinding machine driven by a two-horsepower motor Wood-working machines and machine tools are driven by motors in the carpenter and machine shops In addition to the uses already mentioned, the electric power is also used to pump water, shear the sheep, clip the horses, wash, dry and iron the clothes, heat the house, cook the food, freeze the ice cream, cool the house in the summer, curl the ladies' hair and play the piano

The "Heart's Delight" farm power equipment is much more extensive than would be warranted on a farm of ordinary size, but the installation serves to illustrate the extent to which the application of power may be carried, on an unusually large produce farm In many instances a community of farmers could develop such a water power and distribute the power among themselves to mutual advantage and profit.

DEVELOPING A SMALL WATER POWER

The prime requisite to the creation of a water power is the existence of falling or flowing water. The amount of power which may be available varies, first, with the amount of water flowing, and second, with the amount of fall It requires about one cubic foot of water per second, falling through a height of ten feet, to make available one theoretical horsepower. The fall may be either naturally concentrated at one point in a cascade or it may be artificially concentrated, for the purpose of development, by combining the fall of several cascades or a series of rapids This may be accomplished by either of two methods, first, by building a dam at the downstream end of the rapids to impound the water so that the entire fall is concentrated at the dam, or second, by building a dam at the upstream end of the rapids and conducting the water through a closed pipe to the lower end of the rapids, where the resulting water pressure will be exactly the same as in the first instance. A variation of the latter method consists of diverting the water from the natural channel at the head of the

Cascade on Indian Creek, Warren Co., N. Y. Typical Example of Undeveloped Water Power

rapids and carrying it in a canal, on a slight down grade, along the side of a hill to a suitable point at which the water is turned into penstocks which run directly down the slope to the stream, where the power development may be made. The latter method, involving the construction of a canal, is open to the objection that considerable trouble is usually experienced from the accumulation of ice in the winter time. The first two methods described are the most common.

The amount of water which flows in a stream, in New York State, whether large or small, is subject to remarkable variation. Only one who has observed very carefully and continuously, by actual measurement, the extremes of fluctuation to which a flowing stream is subject, is in a position fully to appreciate this. Some of the larger rivers of New York State are subject to such fluctuations of flow that the amount of water discharged during flood periods is several hundred times as much as the amount that flows in the extreme dry period. Also in many instances from one-half to three-fourths of the total runoff of the stream during the year occurs during a period of a few weeks in the spring months, when the accumulated snow and ice is melted and runs off in conjunction with the warm spring rains. Unfortunately, reliable data relating to the fluctuations of small streams in this State are very meager. It is, however, a matter of record that the smaller streams for which records are available are subject to greater fluctuations per unit of tributary watershed area than are the larger streams. It seems logical, therefore, to assume that the very small creeks and brooks are subject to fluctuations relatively greater than those recorded for streams of only relatively small size. This fact must be borne in mind by any one who proposes to develop the power on a stream, for if it is overlooked the project is not so assured of success. For most purposes power is required in about the same amount for all seasons of the year, while, as previously stated, the streams run off most of their waters in the spring. Therefore, in developing the power of any particular stream, if the power is required to be fairly constant at all seasons of the year as is usually the case, there are two considerations which must not be overlooked.

First — Will the minimum flow of the stream — that is, the flow which occurs in the driest season of a dry year — be sufficient to furnish the amount of power required?

Second — If the minimum flow is not sufficient, what means are available for storing the surplus water from the wet season until the dry season?

The subject of equalizing stream flow throughout the year by means of storage reservoirs has been so thoroughly discussed in the reports of the Commission that further discussion in this connection does not seem warranted

Taking a general average throughout the State of New York, large streams may be depended upon to produce from one-twentieth to one-quarter of a cubic foot of water per second per square mile of tributary drainage area, during the driest period Streams having only one or two square miles of drainage frequently dry up entirely in the dry seasons. If a power development is proposed of such a character that some considerable sacrifice of power might be made in the dry seasons with no serious loss, most small streams may be developed to provide for as much as one-quarter to one-half of a cubic foot per second per square mile On the other hand it is often found practicable to provide a small auxiliary power plant, such as gasolene or kerosene, to fall back upon in dry weather, or to supply extra power occasionally, in which case the water-power development need not be limited to the minimum flow of the stream

The power of falling water may be applied to practical purposes in several ways One of the simplest ways, should it be desired to use the power of the stream to pump water, is by means of what is known as a hydraulic ram This is a device which operates on the principle of the impact due to the sudden stoppage of flow of a column of water By means of this device, or engine, water falling through a very small height may be used to raise a portion of the same, or a comparatively small amount of other water, to an elevation considerably higher than the supply The mechanical efficiency of the hydraulic ram is comparatively high under certain conditions but generally is very low, useful work which manufacturers claim may be realized varying from 38 per cent to 80 per cent The minimum fall under which a ram will effectively elevate water is about two feet This fall will elevate about one-thirteenth of the supply to a height of twenty feet. Under the most favorable conditions and a fair amount of fall, a ram may elevate water as high as 120 feet. The proportion of water which may be elevated varies from one-twentieth to two-sevenths of the total supplied, and, accordingly, the proportion of water which must be wasted at the impetus valve of the ram varies from five-sevenths to nineteen-twentieths These proportions both depend upon the ratio of the amount of supply to the amount to be elevated, that is, a small proportion may be elevated to a considerable height and vice versa In cases where a small brook of suitable quality is available for domestic water supply, it is often entirely practicable to install a hydraulic ram which will pump a sufficient proportion of the amount of supply to furnish a household with all the water necessary for ordinary domestic purposes, in spite of the fact that the brook may be on a lower level than the house Owing to the fact that a hydraulic ram may be applied only to the purposes of elevating water, it is not generally considered as a means of developing water power, although in the broadest sense it does constitute such a development.

On the other hand, the purposes for which power is usually required are not only for the elevation of water for a water supply, but for many other and varied requirements In such cases the power must be developed in such manner that it may be utilized to operate machinery near the site of the development, or transmitted for some distance, and there used to operate machinery or for lighting or heating To develop water power in this manner requires some kind of a water-wheel

There are several types of waterwheels, the principal ones being known as "undershot," "overshot," "breastwheel," "turbine" and "impulse" The overshot wheel is a type very familiar to most readers, being usually of home manufacture It consists, usually, of a wooden wheel with water compartments arranged at regular intervals around the periphery The water is fed into the wheel at the top, just off the center It flows into the compartment at the top and the weight being exerted on one side of the supporting axle causes the wheel to revolve, the water spilling out when the compartment, or water pocket, reaches the bottom This type of wheel depends entirely for its power upon the weight of the water which causes the wheel to revolve

The undershot wheel is very similar in construction to the overshot type but depends more for its power on the velocity of the flowing water which strikes the blades, or buckets, on the under side of the wheel.

The breastwheel is also similar in construction but is in reality an improvement upon the overshot and undershot types. It depends for its power on a combination of the action of gravity and the impulse of the water striking the blades, or buckets. The water is fed into the wheel a little below the height of the axle and usually enters with considerable velocity, a part of which is transformed into useful work by the wheel.

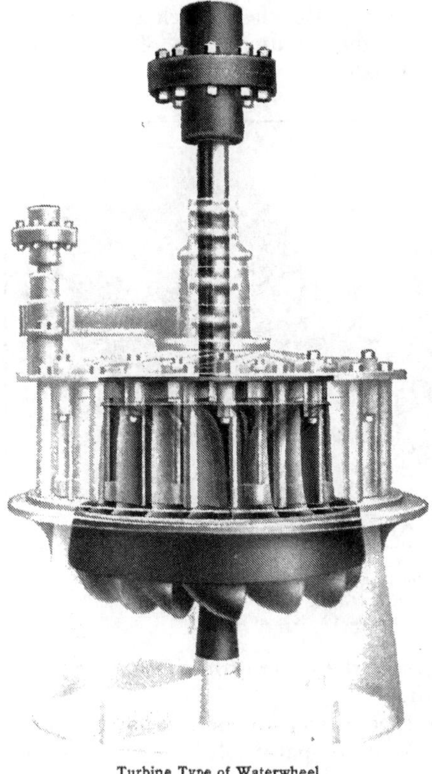

Turbine Type of Waterwheel
Phantom view of wheel-case

The turbine is a type of wheel which is very extensively used. It is usually constructed of metal and consists primarily of a series of curved vanes, or runners, whose arrangement is similar to a screw. The action of the water flowing through these curved vanes causes the vanes and shaft to revolve, the vanes being solidly connected to the shaft, which may be either horizontal or vertical.

The fundamental working principle of an impulse waterwheel is the turning into useful work of the impulse due to the velocity of a jet of water issuing from a contracted orifice. This is accomplished usually by conveying the water from the dam or other source of supply to the waterwheel in a pipe of comparatively large size and then gradually reducing the size of the pipe immediately in front of the wheel to a comparatively small size by means of a reducer section, which is fitted with a nozzle the opening of which may or may not be regulated in size. This contraction of the stream of flowing water causes a spouting of the water under pressure and the water issues in a jet with very high velocity. The jet thus issuing from the nozzle strikes the cups of the impulse wheel which are arranged at regular intervals around the circumference of a metallic disc which is centered on an axle. The cups transfer the veloc-

Impulse Type of Waterwheel
Showing jet of water striking cups. Wheel illustrated is very powerful, but principle of small wheels is the same

ity of the jet to the wheel, and the water drops from them with very little velocity left in it.

In general, the turbine type of wheel is best adapted to low heads, or falls, and the use of comparatively large volumes of water, and the impulse wheel is best adapted to the use of a comparatively high head, or fall, and a comparatively small amount of water. There are certain intermediate conditions for which the manufacturers of each type claim their wheel is best suited and in such instance a study of local conditions is always necessary to determine which type of wheel is best adapted.

The development of a water power by means of any kind of a waterwheel results in the conversion of the energy of the falling water into mechanical power which is exerted in a more or less rapidly revolving shaft. In order to apply this power of the revolving shaft to some useful purpose, there are several methods which may be used. The shaft may be directly connected to the shaft of an electric generator, or dynamo, to generate electric current, or it may be directly connected to a machine which it is desired to operate, provided the machine, or dynamo, is required to operate at the same speed as that of the wheel shaft. This is frequently not the case, so that under ordinary conditions the shaft of the wheel is fitted with a pulley, which in turn is connected by belt to another pulley on the machine which is to be driven.

By using pulleys of different diameters on the shaft of the waterwheel and the shaft of the machinery to be driven, the speed of the machine may be several times more or less than the speed of the waterwheel. For instance, if the waterwheel revolves 200 revolutions per minute and it is desired to operate a machine, connected by belt, at a speed of 1000 revolutions per minute, a pulley of comparatively small size, say four inches in diameter, is placed on the shaft to be driven, and a pulley of five times the diameter, or twenty inches, is placed on the shaft of the waterwheel. This causes the shaft of the machine to revolve at a speed five times as great as the waterwheel. If the speed of the waterwheel is greater than that required for the machinery to be operated, then the reverse operation is followed out, placing a small pulley on the shaft of the waterwheel and a larger one on the shaft of the machinery to be driven. If the speed of the waterwheel is to be magnified more than about six times, it usually requires the installation of a countershaft and another series of pulleys in order to avoid the use of very large and very small pulleys. A pulley which has a very small diameter does not operate satisfactorily without considerable loss of power, and a very large pulley is objectionable on account of the space which it requires.

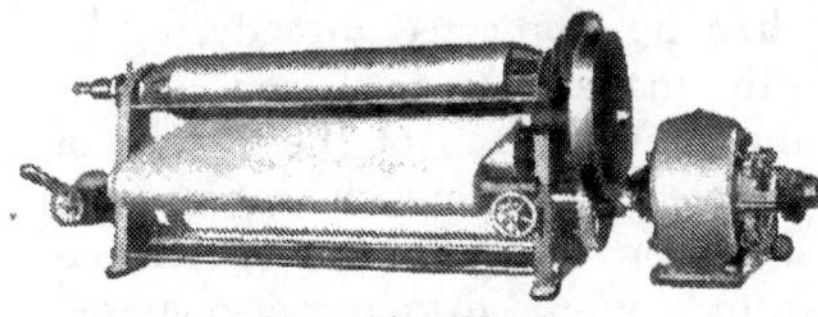

Motor-driven Mangle

When a water power is once developed it may be applied to practical use either near the place of development or at a considerable distance. If it is to be used for power only, and not for lighting, and can be used where it is developed, there is no need of converting it into electricity. But if it is to be used for lighting, or for power to be applied at a considerable distance from the water-power site, then it becomes necessary to convert the power into electricity, in which form it may be most conveniently transmitted from one place to another. This requires an electric generator, or dynamo, to be driven by the waterwheel, and a transmission line, preferably of copper or aluminum wire, to carry the current where it is to be used. In order to reconvert the current into power at the end of the transmission line, where the power is to be used, it is necessary to run the current into an electric motor, the shaft of which is made to revolve by the action of the electric current. This motor may then be connected directly, or by belt, gears or chain drive, to the machine to be driven.

It should be borne in mind that in each of these steps of changing from water power to electric current, in transmitting the current over the wires, in reconverting it into power, and in transferring this power from a motor to a power-operated machine, there are losses of energy. These losses vary considerably in different instances. Assuming, for illustration, that a water power, whose theoretical power is ten horsepower, is required to drive a power machine at a distance, the efficiencies and losses will be somewhat as follows:

Waterwheel,	efficiency	80%,	Loss	20%,	generates	8.0	horsepower
Connections,	"	95%,	"	5%,	transfers	7.6	"
Dynamo,	"	90%,	"	10%,	generates	6.8	"
Transmission,	"	90%,	"	10%,	transmits	6.2	"
Motor,	"	90%	"	10%,	develops	5.5	"
Connections,	"	95%,	"	5%,	delivers	5.0	"

Therefore, only five horsepower would be actually delivered to the machine to be driven. This amounts to only half of the theoretical power of the falling water which is actually realized in useful work of the machine being driven. If the power from the waterwheel is to be applied directly without generating electricity a much higher efficiency will be realized.

ACKNOWLEDGMENT

On behalf of the State Water Supply Commission and the writer, grateful acknowledgment is made to the following named persons who have extended courtesies to me by furnishing information or illustrations for use in connection with the preparation of this pamphlet

Mr E Burdette Miner, Oriskany Falls, N Y
Mr R K Miner, Little Falls, N Y
Mr Jared Van Wagenen, Jr , Lawyersville, N Y
Mr John T McDonald, Delhi, N Y.
Mr Edward R Taylor, Penn Yan, N Y
Mr John Liston, General Electric Company, Schenectady, N Y
Mr R E Strickland, General Electric Company, Schenectady, N Y
Mr Stephen Loines, Brooklyn, N Y
Mr George E Dunham, Utica, N Y
Pelton Water Wheel Company, New York and San Francisco.
James Leffel & Company, Springfield, Ohio

D R COOPER

ALBANY, JANUARY 25, 1911.

PUBLICATIONS OF

STATE WATER SUPPLY COMMISSION

STATE OF NEW YORK

REPORTS

First Annual Report Published February 1, 1906
Includes Commission's annual report on applications for approval of plans for public water supplies, also summarized statistics of public water supplies and sewage disposal in New York State Edition exhausted

Second Annual Report Published February 1 1907
Includes Commission's annual report and decisions on applications for approval of plans for public water supplies, also summarized statistics of public water supplies and sewage disposal in New York State, supplementary to statistics published in First Annual Report, also report on River Improvements for the benefit of public health and safety
Edition exhausted

Third Annual Report Published February 1, 1908
Includes Commission's annual report and decisions on applications for approval of plans for public water supplies, also report on River Improvements for the benefit of public health and safety, also contains Commission's first Progress Report on Water Power and Water Storage Investigations made under chapter 569 of Laws of 1907, including details of Sacandaga and Genesee river studies Edition exhausted

Progress Report on Water Power Development Published March 1, 1908
This is a revised reprint of the part of the Commission's regular Third Annual Report relating to Water Power and Water Storage Investigations showing results of engineering studies up to date of publication

Fourth Annual Report Published February 1, 1909
Includes Commission's annual report and decisions on applications for approval of plans for public water supplies, also report on River Improvements for the benefit of public health and safety, also contains Commission's second Progress Report on Water Power and Water Storage Investigations, with special details of Raquette and Delaware river studies and supplementary studies on Upper Hudson and Genesee, also a census of water power developments in the State

Fifth Annual Report Published February 1, 1910
Includes Commission's annual report and decisions on applications for approval of plans for public water supplies, also summarized statistics relating to public water supplies approved by the Commission in New York State, also report on River Improvements for the benefit of public health and safety, also contains Commission's third Progress Report on Water Power and Water Storage Investigations, with details of reconnaissance studies of Ausable, Saranac, Black, Oswegatchie and other rivers, and a draft of a proposed Water Storage Law

Sixth Annual Report Published February 1, 1911
Includes Commission's annual report and decisions on applications for approval of plans for public water supplies, also report on River Improvements for the benefit of public health and safety, also contains Commission's Fourth Progress Report on Water Power and Water Storage Investigations, with details of investigations of Black and Oswego river watersheds, and a revised draft of a proposed Water Storage Law

MISCELLANEOUS

Pamphlet —" New York State Water Supply Commission " Published September 1909
Issued for distribution at State Fair at Syracuse 1909

Pamphlet —" New York's Water Supply and Its Conservation, Distribution and Uses " Published September, 1910
Issued for distribution at State Fair at Syracuse, 1910

Pamphlet —" Water Resources of the State of New York " Published September 1910
By Henry H Persons, President of the State Water Supply Commission
Issued for distribution at National Conservation Congress at St Paul, Minnesota, 1910

Pamphlet —" Water Power for the Farm and Country Home " Published January, 1911
By David R Cooper, Engineer-Secretary to State Water Supply Commission

Printed in the USA
CPSIA information can be obtained
at www.ICGtesting.com
LVHW022027011023
759825LV00006B/106